ABSTRACT

I0505448

Solar chimney power plant is a technology can be used for power generation. To replaced coal fired power plant one of the suitable options is solar chimney power technology. The solar chimney power technology has three main components named solar collector, power generation unit and chimney. The chimney plays an important role in the solar chimney power generation system. It creates sufficient draft so that turbine can generate electricity. The main identified problem in the solar chimney power generation system is the height of the chimney. To generate 200MW power the estimated chimney height is about 700 to 800 m. Therefore, in this study a model chimney is designed and tested to understand the behavior the chimney. The model chimney was tested under different heat load conditions. To understand the performance of the chimney, the temperatures and flow rate were measured at the different points of the chimney. The chimney was modified with wire mesh screen at the exit. The performance of the modified chimney was also measured and compare with normal chimney. It was found that at higher heat gain ratio modified chimney showed better performance than normal chimney.

ACKNOWLEDGEMENTS

At the outset, I gratefully acknowledge our honorable supervisor Dr. Md. Mizanur Rahman. For his guidance, incessant encouragement and unwavering support. Without his guidance, we would not able to finish this work. My sincere thanks are due to our Head of the Department Farhan Mahbub, Honorable Prof. Dr. S. M Fazlul Karim and all our lecturers in Mechatronics Engineering Department. Furthermore, I would like to express our gratitude to the staffs associated with audio Visual Lab.

This acknowledgment would not be complete without mentioning the invaluable support offered by my friends that helped me overcoming some difficulties encountered during this work. All of your kindness and support means a lot to me. Words are not enough to express my sincere gratitude towards our parents for their unconditional love, devoted support and continuous encouragement throughout the journey.

Thank You
The Authors

TABLE OF CONTENTS

LIST OF TABLES

LIST OF FIGURES

LIST OF ABBREVIATIONS

AC   Alternative Current

LCD   Liquid Crystal Display

LIST OF SYMBOLS

mm	-	Millimetre
V	-	Voltage
s	-	Second
m	-	Minute
V	-	Velocity
°c		Degree Celsius
T	-	Temperature
m/s	-	Metre Per Seconds
I	-	Current
P	-	Power
ρ	-	Density
d	-	Diameter
A	-	Area
M	-	Mass
C_p	-	Specific Heat Capacity
Q	-	Specific Heat
ΔT	-	Temperature Difference
Kg	-	Kilogram
K	-	Kelvin
kJ/kg.K	-	Kilo Joule Per Kilogram Kelvin
J/Kg	-	Joule per Kilogram
m^2	-	Square Metre
kg/m³	-	Kilogram Per Cubic Metre

CHAPTER 1

INTRODUCTION

CHAPTER 1
INTRODUCTION

The future of this earth and mankind substantially depends on our ability to slow down the population increase in the "Third World" by civilized means. The key is to increase the standard of living, to overcome the inhumane poverty and deprivation. Development requires mechanization and energy. Energy consumption increases proportionally to the gross national product or prosperity while simultaneously the population growth will decrease exponentially. Many developing countries possess hardly any energy sources and their population doubles every 15- 30 years. The result of which are commonly known that is civil wars and fundamentalism. If these developing countries are provided with only a humane and viable minimum of energy the global energy consumption will drastically increase. The sun is the only source which can supply such an enormous amount of energy without an ecological breakdown and without safety hazards and without a rapid depletion of natural resources at the expenses of future generations. Solar chimneys are such devices which can generate energy up to large extent by use of the simple device that is greenhouse collector, Vertical chimney and turbine and produce electricity continuously at minimum cost. It is very necessary to adopt the solar technology in every part of the world like other conventional sources which we are using regularly. [1]

1.1 Background

The natural draft cooling tower has long been associated with thermal power plants to discharge waste heat to the atmosphere, and nuclear power plants are no exception. Nowadays, with the growing environmental concern this type of tower is more and more considered as a valuable solution due to its quiet operation, longevity and its ease in maintenance and fuel savings. With capacity reaching hundreds of megawatts, any under-performance of cooling tower means additional cooling equipment for the cooling water to the turbine condenser to ensure the thermal efficiency is maintained.

The technology combines three components: a collector, a chimney and turbines. In the collector, solar radiation is used to heat an absorber (ordinarily soil or water bags) on the ground, and then a large body of air, heated by the absorber, rises up the chimney, due to the density difference of air between the chimney base and the surroundings. The rising air drives large turbines installed at the chimney base to generate electricity. The concept of solar chimney power technology was first conceived many years ago, again presented in 1978 and proven with the operation of a pilot 50 kW power plant in Manzanares, Spain in the early 1980s.

The solar collector is made of a transparent cover, raised to a certain height from ground level, a storage system which might be the ground located underneath the transparent cover to which a tub filled with water can be added in order to ensure continuous and uninterrupted operation of the plant for a longer period. Figure 1.1, The chimney is located at the center of the collector and the turbine is positioned at its base. Basically, solar radiation passing through the transparent cover hits the collector absorber which might be the ground or the tub upper surface. Once heated, the absorber transfers some heat to the inside air of the collector. Thus, the air density differences make the air move inside the chimney resulting in an air flow from the collector entrance to the chimney outlet. The kinetic energy of the fluid is converted into electrical energy through wind turbines placed at the entrance of the chimney.

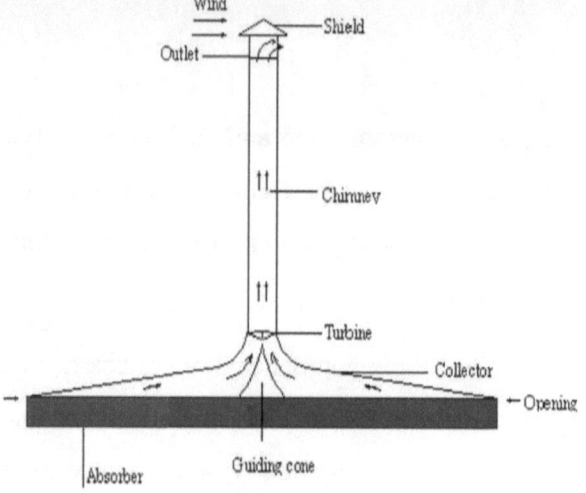

Fig 1.1: Schematic diagram of solar chimney power generating system.

The low energy densities and unstable discontinuous power output prevent wind and solar from being a primary source of energy. Mechanisms to mitigate the spatiotemporal intermittency of the wind or to overcome the limitation of the sunshine duration are needed. This system utilizes a basic principle of physics, that "hot air rises," to generate wind. This then induces a thermal updraft inside the cylindrical tower placed at the center of the solar collector. This thermal updraft turns the wind turbine located at the base of the tower in order to produce electricity.

1.2   Problem Statement

The physical parameters of the solar chimney play important roles in enhancing draft, other than external factor such as solar irradiation, air temperature, cross wind, outdoor wind direction, wind velocity and so forth. The ability of solar chimney generated power is depending on the air speed inside the solar chimney. Number research has been conducted on solar chimney length and solar collector diameter gap in order to enhance the performance of solar chimney to generate electricity. Cold inflow and turbulence would happen if the solar chimney diameter was too big. The cold inflow or downward flow of air is observed at the center of the chimney. The entire airflow field of the solar chimney power plant is actually under the joint influences of the buoyancy force, and the crosswind effects at the solar collector entrance, and the chimney outlet. The crosswind influences the path of the air intake at the solar collector. The study was also reported that air flow rate increased from 0.001 m^3/s to 0.03 m^3/s when the lengths of the solar chimney increased from 0.05 m to 3 m. A recent Computational Fluid Dynamics (CFD) study has be conducted by Ming et al. on the influence of ambient crosswind on the performance of a solar chimney power plant. The ambient cross wind has positive and negative effects on the solar chimney power plant performance. At weak ambient cross wind deteriorated the flow field and reduce performance significantly but the strong crosswind enhance the mass flow rate results output power increased. The strong ambient cross wind significantly enhanced the wind suction effect at the top of the chimney. In addition, the crosswinds influence the performance of the solar chimney by convective heat losses from the collector surface to the environment. The cross wind is dragging hot air out from the solar collector and reduce the collector efficiency. To overcome the negative effect of cross wind side wall or side blockage a few meter away is a suitable option. It was also found the cross wind magnitude on the Beaufort scale (8 ms^{-1} to 17 ms^{-1}) significantly reduce the performance of the chimney is about 10 to 25%. It is found that cold air is liable to 'sink' into the glass tube or a square duct that like a chimney under typical exit bulk velocities, of 3 ms-1 to 5 ms-1 or below in a quiescent ambient condition

1.3   Objectives

i.   To reduce the cold inflow effects at the chimney, exit for restoring the energy losses

ii.   To understand the behavior of chimney under different heat load

iii. To determine the effect of wire mesh screen on performance of the chimney

iv.   To develop an empirical formula that explains behavior of the chimney.

1.4   Overview

Chapter 1 introduces the background of the Solar Chimney power generation technology & the aims and the objectives of the project.

Chapter 2 describes the solar chimney energy produce process and other literature reviews related to the project.

Chapter 3 explains the procedure of designing the Chimney & explains how we connected each and every components part to make the project.

Chapter 4 shows the outcome of the project and discusses the results.

Chapter 5 finds out the weak points of the project and the ways to improve.

CHAPTER 2

LITERATURE REVIEW

7

CHAPTER 2
LITERATURE REVIEW

Researchers and engineers developed the solar chimney power generation technology using wind turbine over the past few decades.

Xinping Zhou, Jiakuan Yang, Bo Xiao, Guoxiang Hou, Fang Xing et.al (2009) experimented that the maximum chimney height for convection avoiding negative buoyancy at the latter chimney and the optimal chimney height for maximum power output are presented and analyzed using a theoretical model validated with the measurements of the only one prototype in Manzanares. Sensitivity analyses are also performed to examine the influence of various lapse rates of atmospheric temperatures and collector radii on maximum height of chimney. [1]

Xinping Zhou, Fang Wanga,b, Reccab M. Ochieng et.al (2010) presented that a comprehensive picture of research and development of SC power technology in the past few decades & described the description, physical process, experimental and theoretical study status, and economics for the conventional SC power technology are included as well as descriptions of other types of SC power technology. [2]

Chi-Ming Chu, Md. Mizanur Rahman, Sivakumar Kumaresan et.al (2011) experimented that Temperature and pressure drop data obtained from an air-cooled heat exchanger model with cross-sectional flow areas operating under natural convection are presented that indicate significant cold inflow, resulting in the reduction of effective chimney height. Additional tests on smaller scale models appeared to favor the explanation that the occurrence of cold inflow in the air-cooled heat exchanger model was primarily due to the relative ease in either drawing cold air from inlet or from outlet. [3]

8

C. Chi-Ming Chu, R. Kwok-How Chu & M. M. Rahman et.al (2012) experimented that the cold air inflow into a natural draft air cooled heat exchanger chimney. It was found that the Top-only, Top-and-Middle, and Middle-only installations were able to maintain draft velocities to within 3.0 % of the control experiment draft velocity, in spite of the resistance of the mesh & also found that configurations utilizing mesh at the bottom had experienced a drop in draft velocity by up to 50% from the Control. The smoke flow visualization tests clearly proved that cold air inflow phenomenon exist, and can be countered with the installation of wire mesh at the top of the model. [4]

Amel Dhahri, Ahmed Omri et.al (2013) presented that a technology of electric power generation using solar energy by employing basic physics that when air is heated it rises. The created updraft can be used to turn a turbine placed at an appropriate position within a tall chimney to generate electricity. [5]

P. J. Bansod, S. B. Thakre, N. A Wankhade (2014) presented that the critical review of this important technology in the form of the advancements and developments taken place in various parts of the world and analyzes its important aspects. [6]

Shinsuke Okada, Takanori Uchida, Takashi Karasudani, Takashi Karasudani(2015) experimented that the mechanism in order to augment the velocity of the air which flows into the wind turbine by applying a diffuser tower instead of a cylindrical one, the efficiency of the systems power generation is increased. The mechanism that we investigated was the effect of the diffuser on the solar chimney structure. The inner diameter of the tower expands as the height increases so that the static pressure recovery effect of the diffuser causes a low static pressure region to form at the bottom of the tower. This effect induces greater airflow within the tower.[7]

Hakim Semai, Amor Bouhdjar & Salah Larbi (2016) states that the modeling of turbulent flow under the effect of natural convection within a solar chimney power plant (SCPP) by performing numerical simulation using the Saturne Code coupled with Syrthes code. The concept of minimizing the entropy production is also studied with the objective of optimizing the geometric configuration as well as the effect of the collector cover slope on the efficiency of SCPP. It is focus on the storage system influence on the SCPP performance and the duration of its operation after sunset. This leads to the improvement of the global efficiency of the SCPP and the positive impact of the extra storage media use and the configuration which improves the velocity at the chimney entrance. [8]

Amin Mohamed El-Ghonemy et.al (2016) said that the solar chimney power plant (SCPP) is a large scale power plant & is applicable in desert areas, where solar radiation is good. Using a larger solar collector diameter, a greater volume of air is warmed to flow up in a higher chimney & also showed that, the solar chimney power plant, in which the chimney height and diameter are 200 m and 10 m, respectively, and the diameter of the solar collector diameter is 500 m, can produce a monthly average of 118~224 kW electric power during all the year. [9]

W. M.A Elmagid, I. Keppler et.al (2017) analysied that the redesign of an axial turbine for SCPP of Manzanares prototype is presented to simulate SCPP overall with using radiation model. Additionally, the investigation of flow inside the turbine is carried out by using the three dimensional CFD model. The CFD model solves Reynolds-averaged Navier–Stokes equations (RANS equation) using K-ε turbulent model. The solar radiation is calculated by using two different radiation models according to the physical state. The comparison of the CFD results and previous experimental results show a good agreement, which validates our CFD model. [10]

Omer Khalil Ahmed, Abdullah Sabah Hussein et.al (2018) shows that two experimental models of a hybrid solar chimney were built and designed (systems A&B) & results showed that system (A) had higher thermal gain than system B while the daily average of electrical power in system (B) was (75.6 W) higher than system (A) (79 W). This is because the high thermal gain raised the operating temperature of the PV panel which led to a decrease in its power output. The total useful power produced by the system (B) is greater than the useful power produced from the system (A). [11]

Qingjun Liu, Fei Cao, Yanhua Liu, Tianyu Zhu, and Deyou Liu (2018) experimented that the Performances of the designed SCPVTPP are then simulated. The SCPVTPPs with different PV module areas are finally discussed. It is found that the PV cells hold the highest temperature in the solar collector. Temperature rise of the PV module has significant influences to its power generation. With the increase of the solar collector ratio, the temperature rise and the wind velocity both first decrease then increase, the SCPP power productivity decreases linearly, and the PV power productivity increases linearly, whereas the PVT power productivity first increases linearly then increases super linearly. There is a reversed solar collector ratio, exceeding which the PV generates most power. [12]

Ajay Bejalwar, Pramod Belkhode (2018) presented that the formulation of mathematical model for experimental setup of a small chimney power plant. The base of collector is covered with sand on the ground layer of brick and concrete. [13]

CHAPTER 3

METHODOLOGY

CHAPTER 3
METHODOLOGY

For the purpose of development of the solar chimney energy generation, a prototype system has been developed. By establishing the various variables of the system to improvising the energy generation process.

3.1 Block Diagram

A voltage regulating system is kind of operation that is controlling voltage which is supplied to the system. In voltage regulator is connected to the electric heater to heat the air. This air is inflow the chimney & heated inside the chimney & air flow is increases with time. To measure the supply voltage & current uses voltmeter in parallel & ammeter in series to the voltage regulator. Here we uses two temperature sensor to measure inside the chimney temperature. First temperature is used in middle of the chimney (which is used 10.4 inchs below of the chimney top) & second temperature sensor is used in top of the chimney (which is used 0.45 inchs below of the chimney top). Temperature sensor is used for measuring the whole chimney temperature. In chimney we used a mesh top of the chimney for reducing the cold inflow effect which is reduce the efficiency of energy generation.

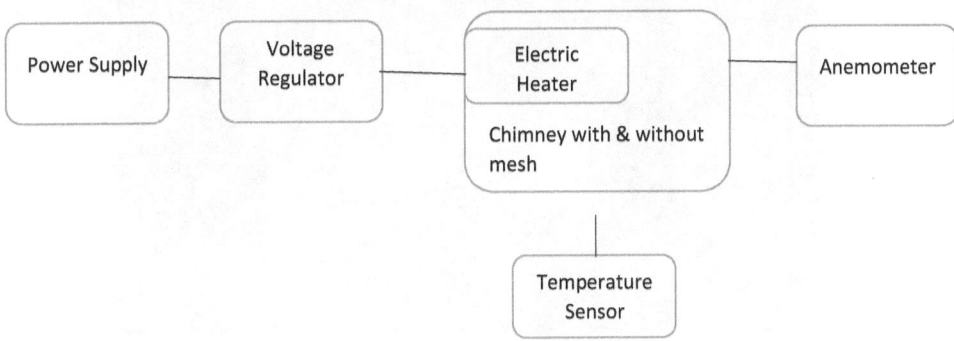

Figure 3.1: Block diagram of solar chimney power generation technology

3.2 Mechanical Design

To develop the mechanical design of this work, we have used two temperature sensor in different places to measure the temperature to the chimney. And also used mesh top of the chimney to remove the cold inflow effect which is reduce the energy produced efficiency. In figure 3.2 shows the mechanical design of solar chimney energy produced technology without effect of inflow of cold air.

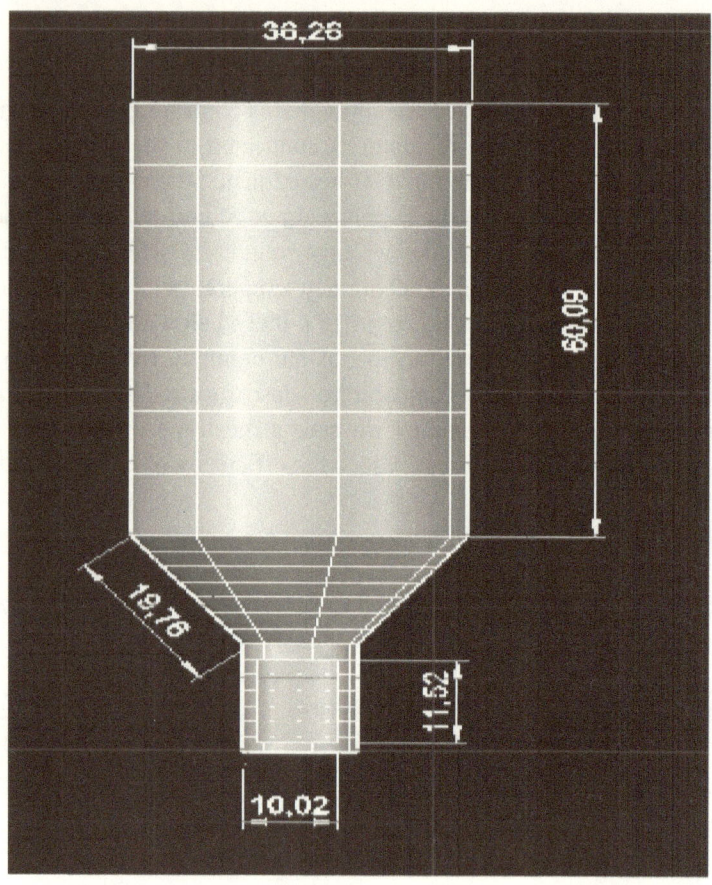

Figure 3.2: Mechanical design of Solar Chimney

A chimney is used to produced energy using the natural air to rotates the turbine & generate energy. Which is using the air flow of the chimney & heated the air to increases the air velocity. Also uses the mesh top of the chimney to remove the cold air flow effect.

The mechanical section of this project is designed with following components which are shown in table 3.1.

Table 3.1: Mechanical Components

COMPONENTS	QUANTITY	DESCRIPTION
Chimney	01	Used to heating air in a closed area & to get desire velocity
Mesh	01	Used to remove the cold inflow effect

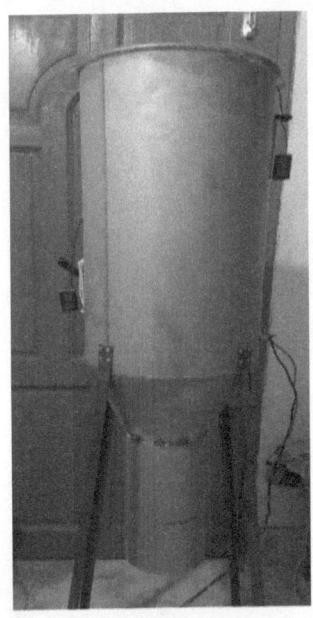

Figure 3.3: Side view of Chimney

Figure 3.4: Top view of Chimney

3.3 Electrical Design

Figure 3.4 shows the electrical circuit diagram of the voltage regulating process to the electric heater. In this figure connected between voltage regulator to the electric heater & also added multi-meter to measure supply voltage & current. Here voltage regulator is supplied at 220V AC.

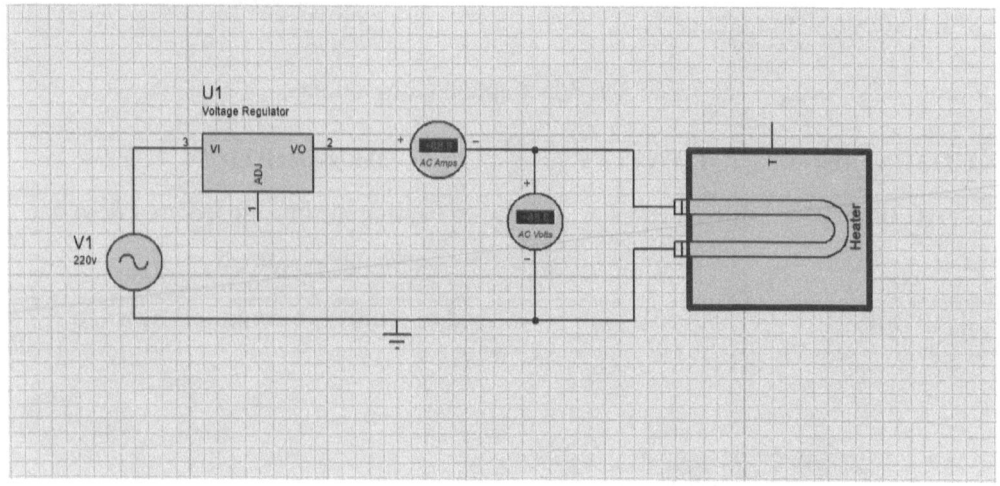

Figure 3.5: Electrical circuit diagram of the system

The electrical section of this work is designed with the following components which are shown in table 3.2.

Components	Quantity	Description
Electric Heater	01	Used to heat the air which is flow in chimney.
Voltage Regulator	01	Used to regulate the supply voltage which is given to electric heater.
Temperature Sensor	02	Used to measure the chimney temperature.
Anemometer	01	Used to measure the air velocity which is inlet the chimney & also measure the out-side temperature of the chimney.

Table 3.2: Electrical Components

Pictures and the details of the electrical products used in this project are given below:

In this work we used 1.5 KW Tubular air heater which is heating the flowing air in the chimney. In tubular heater provide a highly efficient, reliable and cost effective means of supplying heat directly to air, various types of liquids and metals. It is available in straight lengths or bent as per user's requirement with threaded stud connection at both ends. Tubular heaters have enlarged surface area to help faster dissipation of heat. The tubular air heater is shown in below at figure 3.6.

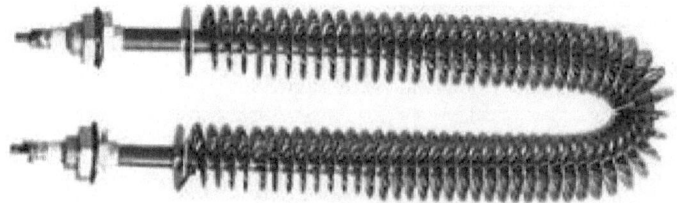

Figure 3.6: Electric air heater

Voltage regulator is used to controlling the suppling voltage. In here we used 1 KW voltage regulator which is supplied maximum 500 volts or 1 KW. A voltage controller, also called an AC voltage controller or AC regulator in an electronic module based on either thyristors, TRIACs, SCRs or IGBTs, which converts a fixed voltage, fixed frequency alternating current (AC) electrical input supply to obtain variable voltage in output delivered to a resistive load. This varied voltage output is used for dimming street lights, varying heating temperatures in homes or industry, speed control of fans and winding machines and many other applications, in a similar fashion to an autotransformer. Voltage controller modules come under the purview of power electronics. Because they are low-maintenance and very efficient, voltage controllers have largely replaced such modules as magnetic amplifiers and saturable reactors in industrial use. In figure 3.5 shows the uses AC voltage regulator.

Figure 3.7: AC Voltage Regulator

Temperature sensor is used to measure the chimney temperature. In this work we used a common thermometer can be used in many case, it has a temperature sensing probe combined with a display, temperature range from -50 ~ 110 degree, display digits are 4 and half, e.g. 102.5, first digit is 1, data flashes in every two seconds. In figure 3.7 shows the temperature sensor.

Figure 3.8: Temperature Sensor

Anemometer is used for measure the inflow velocity of the chimney. It is a mini wind speed and temperature tester. The light weight device is equipped with the latest magnetic sensing technology, which can directly display the air flow speed on LCD. It has various units switching like as m/s, km/h, ft/min, knots, mph. Its range to measuring velocity at 0-30 m/s & temperature is measure at 14-118 ° C. In below figure 3.8 shows the anemometer.

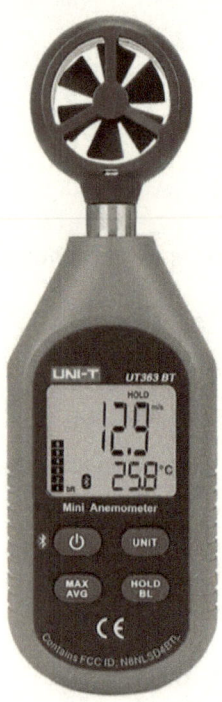

Figure 3.9: Anemometer

3.4 Flow Chart of Solar Chimney Power Generation Technology

When we supply a fixed voltage then the heater is get hot then the air flow is continuously increases to the chimney. At the same time temperature sensor which placed at inside of chimney indicate the internal temperature of air. Another temperature sensor is used for measuring the outside temperature of chimney. After supplying a fixed voltage, we wait for 20 minute so that the air can be heated properly. Then we measured the velocity with an anemometer from the bottom of chimney where turbine can be placed for producing electricity. We take the measurement of inside and outside temperature of chimney and also the velocity of air inside the chimney at the same time after 5 minute. We continue the process till 60 minute and collect the data. We take 5 difference voltage for more correct data analysis and also measured the value of current for those voltage. we installed mesh in the chimney for reduce the cold inflow and increase the efficiency of chimney. Then we run the same experiment with mesh and collect data for different variable. We compare both data with mesh and without mesh to analysis how much efficiency can increases or decreases. After data analysis we complete report and stop our thesis work.

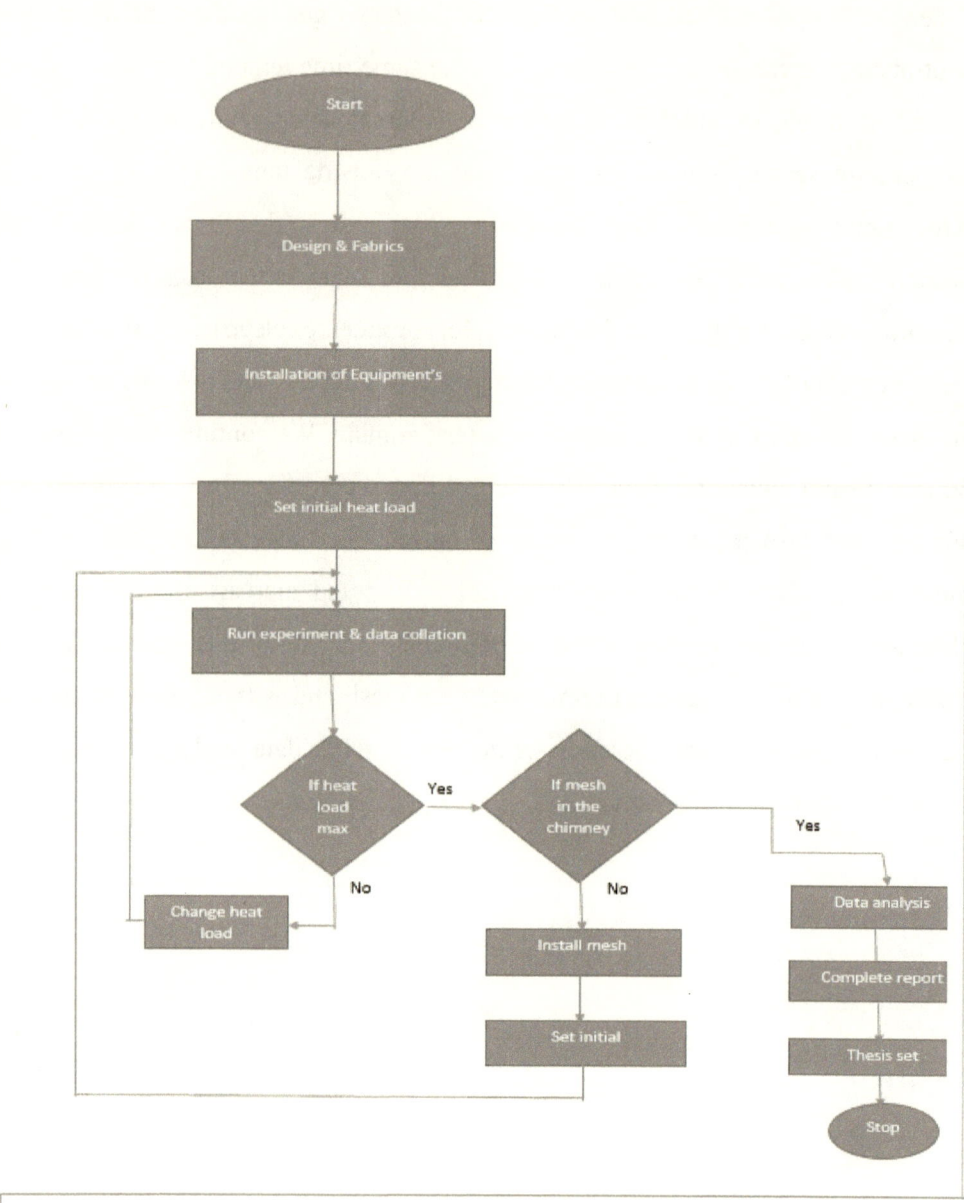

Figure 3.10: Flow chart of Solar Chimney Power Generation Technology

CHAPTER 4

RESULT & DISCUSSION

CHAPTER 4
RESULT AND DISCUSSION

4.1 Result

The temperatures are measured in three different point namely at chimney entrance and exit and middle of the chimney and presented in the Table 4.1. The inlet air temperatures are varied from 31^0C to 40.93^0C and the temperature in the solar chimney is also varied from 52^0C to 96^0C whereas the temperature at the exit of the chimney varied 53^0C to 81^0C. The temperatures at the chimney inlet and outlet and above the heater depend on the voltage and current input in the system. The heat load was 123.75 W at the initial condition and it increase up to 412 W.

Table 4.1: Experimental data of solar chimney

Trend	Voltage (V) Volts	T1 °c	T2 °c	T3 °c	Velocity (v) m/s	Current (I) amp	Power (P) Watt
Without Mesh	150	31.052	51.819	53.28	1.004	0.825	123.75
	175	31.77	61.681	60.42	1.122	0.963	168.525
	200	31.35	70.959	65.15	1.196	1.1	220
	225	41.39	88.867	81.29	1.367	1.238	278.55
	250	40.93	96.026	83.33	1.5	1.65	412.5
With Mesh	150	32.61	50.96	52.6	1.172	0.825	123.75
	175	32.64	59.1	62.36	1.307	0.963	168.525
	200	33	63.88	65.52	1.489	1.1	220
	225	36.35	86.65	90.85	1.218	1.238	278.55
	250	37.66	98.52	98.22	1.396	1.65	412.5

The chimney model is covered with a wire mesh screen and temperatures are measured for different heat load. The highest temperature 98.52⁰C is observed at the middle of the chimney when the chimney exit is covered with a wire mesh screen. This is because the wire mesh screen trapped the heat inside the chimney and enhanced the temperature. According to Chu et al. the wire mesh screen also protracts to penetrate the cold air from outside in known as cold inflow. The cold air does not get chance to meet with the hot air at and reduce temperature.

The temperature for different points in condition of without mesh for power 123.75 W are shown in table 4.2:

The temperatures are measured in three different point namely at chimney entrance and exit and middle of the chimney and presented in the Table 4.2. The inlet air temperatures are varied from 30⁰C to 30.67⁰C and the temperature in the solar chimney is also varied from 50.7⁰C to 52.87⁰C whereas the temperature at the exit of the chimney varied 50⁰C to 55.5⁰C. The temperatures at the chimney inlet and outlet and above the heater depend on the voltage and current input in the system.

Table 4.2: Experimental data of temperature & velocity of air at 150volt supply without mesh condition

Voltage (Volts)		20m	25m	30m	35m	40m	45m	50m	55m	60m
150	T1 (°c)	32.03	32.4	31.33	30.43	31.03	31.27	30.67	30.3	30
	T2 (°c)	50.7	52.77	52.33	51.6	51.9	51.47	51.67	51.07	52.87

T3 (°c)	50	53.33	55.5	54.33	53.67	54.33	52.67	53.33	52.33
Velocity (m/s)	0.933	1	1.033	1	1	1.033	1.067	0.933	1.033

The temperature for different points in condition of without mesh for power 168.525 W are shown in table 4.3:

The temperatures are measured in three different point namely at chimney entrance and exit and middle of the chimney and presented in the Table 4.3. The inlet air temperatures are varied from 31.23°C to 32.83°C and the temperature in the solar chimney is also varied from 57.9°C to 64.5°C whereas the temperature at the exit of the chimney varied 59.9°C to 61.17°C. The temperatures at the chimney inlet and outlet and above the heater depend on the voltage and current input in the system.

Table 4.3: Experimental data of temperature & velocity of air at 175volt supply without mesh condition

Voltage (Volts)		20m	25m	30m	35m	40m	45m	50m	55m	60m
175	T1 (°c)	32.83	32.83	31.57	31.33	31.3	31.23	31.23	31.37	32.23
	T2 (°c)	57.9	61.3	64.5	63.93	61.7	62.63	61.5	60.87	60.8
	T3 (°c)	60.17	61.17	60.53	59.9	60.03	60.33	60.7	60.93	60
	Velocity (m/s)	1.233	1.067	1.133	1.067	1.167	1.1	1.067	1.1	1.167

The temperature for different points in condition of without mesh for power 220 W are shown in table 4.4:

The temperatures are measured in three different point namely at chimney entrance and exit and middle of the chimney and presented in the Table 4.4. The inlet air temperatures are varied from 30.97^0C to 31.77^0C and the temperature in the solar chimney is also varied from 68.9^0C to 72.07^0C whereas the temperature at the exit of the chimney varied 63.97^0C to 65.97^0C. The temperatures at the chimney inlet and outlet and above the heater depend on the voltage and current input in the system.

Table 4.4: Experimental data of temperature & velocity of air at 200volt supply without mesh condition

Voltage (Volts)		20m	25m	30m	35m	40m	45m	50m	55m	60m
200	T1 (°c)	31.03	31.7	31.27	31.5	30.97	31.37	31.1	31.77	31.43
	T2 (°c)	70.53	69.97	71.47	70.9	72.07	68.9	71.43	71.87	71.5
	T3 (°c)	64.5	64.53	64.47	63.97	65.87	65.57	65.77	65.7	65.97
	Velocity (m/s)	1.4	1.233	1.267	1.067	1.167	1.133	1.233	1.167	1.1

The temperature for different points in condition of without mesh for power 278.55 W are shown in table 4.5:

The temperatures are measured in three different point namely at chimney entrance and exit and middle of the chimney and presented in the Table 4.5. The inlet air temperatures are varied from 37.73⁰C to 44.17⁰C and the temperature in the solar chimney is also varied from 85.43⁰C to 92.9⁰C whereas the temperature at the exit of the chimney varied 77.3⁰C to 82.63⁰C. The temperatures at the chimney inlet and outlet and above the heater depend on the voltage and current input in the system.

Table 4.5: Experimental data of temperature & velocity of air at 225volt supply without mesh condition

Voltage (Volts)		20m	25m	30m	35m	40m	45m	50m	55m	60m
225	T1 (°c)	37.73	40.27	40.4	41.63	44.17	41.9	42.33	42.3	41.8
	T2 (°c)	85.43	87.83	92.9	89.17	87.03	87.8	89	89.17	91.47
	T3 (°c)	77.3	80.5	82.63	81.9	80.8	83.0	80.87	80.27	81.28
	Velocity (m/s)	1.4	1.467	1.4	1.467	1.333	1.433	1.267	1.567	1.367

The temperature for different points in condition of without mesh for power 412.5 W are shown in table 4.6:

The temperatures are measured in three different point namely at chimney entrance and exit and middle of the chimney and presented in the Table 4.6. The inlet air temperatures are varied from 36.67⁰C to 42.97⁰C and the temperature in the solar chimney is also varied from 91.17⁰C to 105.3⁰C whereas the temperature at the exit of the chimney varied 76.17⁰C to 90.87⁰C. The temperatures at the chimney inlet and outlet and above the heater depend on the voltage and current input in the system.

Table 4.6: Experimental data of temperature & velocity of air at 250volt supply without mesh condition

Voltage (Volts)		20m	25m	30m	35m	40m	45m	50m	55m	60m
250	T1 (°c)	36.67	42.97	41.3	40.3	40.2	41.3	40.7	42.3	42.7
	T2 (°c)	91.17	99.8	93.07	89.97	92.03	96.87	96.93	99.1	105.3
	T3 (°c)	81.27	86.97	77.53	76.17	81	83.87	85.77	86.5	90.87
	Velocity (m/s)	1.5	1.4	1.4	1.533	1.433	1.6	1.433	1.467	1.733

The temperature for different points in condition of with mesh for power 123.75 W are shown in table 4.7:

The temperatures are measured in three different point namely at chimney entrance and exit and middle of the chimney and presented in the Table 4.7. The inlet air temperatures are varied from 31.8⁰C to 32.97⁰C and the temperature in the solar chimney is also varied from 49.97⁰C to 51.83⁰Cwhereas the temperature at the exit of the chimney varied 50.80C to 54.030C. The temperatures at the chimney inlet and outlet and above the heater depend on the voltage and current input in the system.

Table 4.7: Experimental data of temperature & velocity of air at 150volt
supply with mesh condition

Voltage (Volts)		20m	25m	30m	35m	40m	45m	50m	55m	60m
150	T1 (°c)	32.97	32.67	32.83	32.57	32.97	31.8	32.47	32.57	32.7
	T2 (°c)	50	49.97	51.07	51.5	51.83	50.33	50.93	51.27	51.73
	T3 (°c)	50.8	51.77	53.23	54.03	53.1	51.2	52.37	53.67	53.23
	Velocity (m/s)	1.4	0.933	1	1.3	1.333	1.167	1.1	1.167	1.133

The temperature for different points in condition of with mesh for power 168.525 W are shown in table 4.8:

The temperatures are measured in three different point namely at chimney entrance and exit and middle of the chimney and presented in the Table 4.8. The inlet air temperatures are varied from 32.03°C to 33.10°C and the temperature in the solar chimney is also varied from 56.57°C to 61.7°C whereas the temperature at the exit of the chimney varied 57.3°C to 64.77°C. The temperatures at the chimney inlet and outlet and above the heater depend on the voltage and current input in the system

Table 4.8: Experimental data of temperature & velocity of air at 175volt
supply with mesh condition

Voltage (Volts)		20m	25m	30m	35m	40m	45m	50m	55m	60m
175	T1 (°c)	33.1	33.03	32.57	32.63	32.13	32.03	32.7	32.83	32.73
	T2 (°c)	59.43	59.27	59.27	59.2	58.57	60.17	57.8	61.7	56.57
	T3 (°c)	63.27	63	63.4	62.73	62.83	63.8	60.13	64.77	57.3
	Velocity (m/s)	1.367	1.233	1.1	1.367	1.433	1.233	1.3	1.333	1.4

The temperature for different points in condition of with mesh for power 220 W
are shown in table 4.9:

The temperatures are measured in three different point namely at chimney entrance
and exit and middle of the chimney and presented in the Table 4.9. The inlet air
temperatures are varied from 32.37^0C to 33.47^0C and the temperature in the solar
chimney is also varied from 61.5^0C to 65.77^0C whereas the temperature at the exit
of the chimney varied 61.6^0C to 70.9^0C. The temperatures at the chimney inlet and
outlet and above the heater depend on the voltage and current input in the system

Table 4.9: Experimental data of temperature & velocity of air at 200volt supply with mesh condition

Voltage (Volts)		20m	25m	30m	35m	40m	45m	50m	55m	60m
200	T1 (°c)	32.7	34.1	32.4	32.63	32.37	32.83	33.47	33.17	33.33
	T2 (°c)	65.5	65.77	61.57	65.7	63.97	65.43	63.87	61.63	61.5
	T3 (°c)	70.9	67.97	61.6	65.03	62.8	67.77	66.53	61.6	65.43
	Velocity (m/s)	1.467	1.267	1.4	1.6	1.533	1.433	1.467	1.533	1.7

The temperature for different points in condition of with mesh for power 278.75 W are shown in table 4.10:

The temperatures are measured in three different point namely at chimney entrance and exit and middle of the chimney and presented in the Table 4.10. The inlet air temperatures are varied from 35.73^0C to 36.83^0C and the temperature in the solar chimney is also varied from 84.0^0C to 89.63^0C whereas the temperature at the exit of the chimney varied 88.4^0C to 92.17^0C. The temperatures at the chimney inlet and outlet and above the heater depend on the voltage and current input in the system

Table 4.10: Experimental data of temperature & velocity of air at 225volt supply with mesh condition

Voltage (Volts)		20m	25m	30m	35m	40m	45m	50m	55m	60m
225	T1 (°c)	36.63	36.03	36.83	36.37	36.4	36.67	36.07	36.43	35.73
	T2 (°c)	86.17	88.03	88.1	84.63	84.93	84	87.5	86.87	89.63
	T3 (°c)	88.4	91.53	91.83	89.97	90.77	90.73	90.6	91.63	92.17
	Velocity (m/s)	1.233	1.333	1.267	1.233	1	1.267	1.167	1.267	1.2

The temperature for different points in condition of with mesh for power 412.5 W are shown in table 4.11:

The temperatures are measured in three different point namely at chimney entrance and exit and middle of the chimney and presented in the Table 4.11. The inlet air temperatures are varied from 34.73°C to 40.93°C and the temperature in the solar chimney is also varied from 95.7°C to 100.5°C whereas the temperature at the exit of the chimney varied 97.07°C to 99.97°C. The temperatures at the chimney inlet and outlet and above the heater depend on the voltage and current input in the system

Table 4.11: Experimental data of temperature & velocity of air at 250volt supply with mesh condition

Voltage (Volts)		20m	25m	30m	35m	40m	45m	50m	55m	60m
250	T1 (°c)	34.73	36.43	37.1	37.33	37.57	37.6	37.87	39.37	40.93
	T2 (°c)	97.97	97.23	100.8	98.33	95.7	96.47	99.9	100.5	99.8
	T3 (°c)	99.2	97.97	99.7	97.77	97.33	97.5	97.07	99.03	98.43
	Velocity (m/s)	1.467	1.367	1.467	1.333	1.367	1.367	1.367	1.367	1.467

Analysis

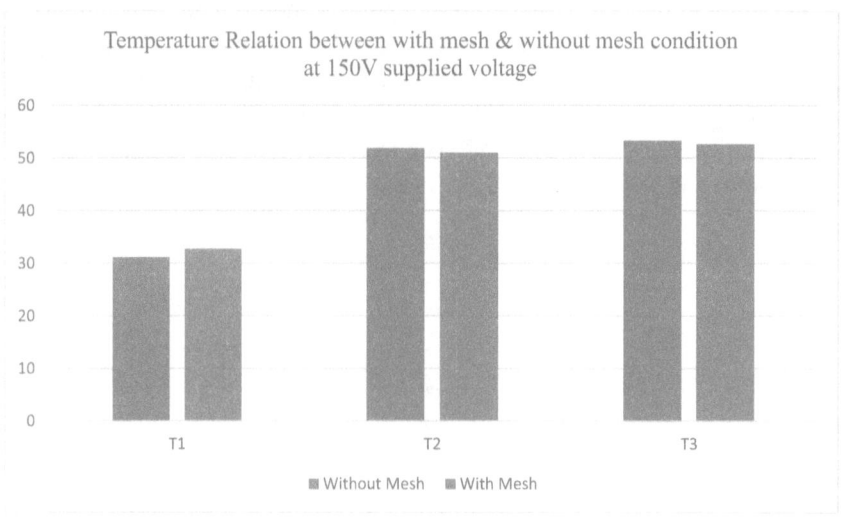

Figure 4.1: Temperature Relation between with mesh & without mesh condition at 150V supplied voltage

In figure 4.1,

For T1 temperature with mesh condition temperature is high cause is atmospheric temperature fall in without mesh condition.

For T2 temperature without mesh condition temperature is high cause of

For T3 temperature without mesh condition temperature is high cause of there is no cross air effect in without mesh condition. In another side with mesh condition have little amount of cross air effect which is known as cols inflow effect.

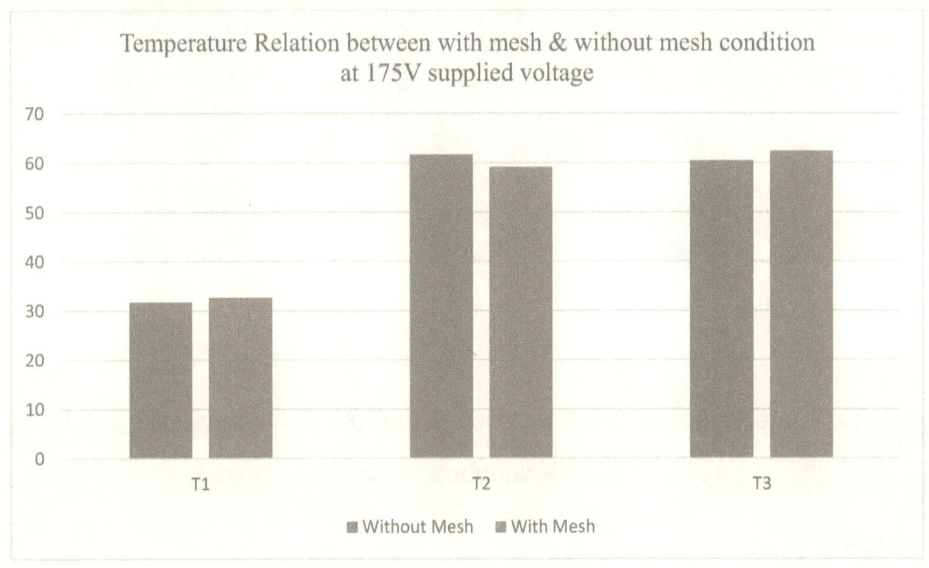

Figure 4.2: Temperature Relation between with mesh & without mesh condition at 175V supplied voltage

In figure 4.2,

For T1 temperature with mesh condition temperature is high cause is atmospheric temperature fall in without mesh condition.

For T2 temperature without mesh temperature is high cause of

For T3 temperature with mesh temperature is high cause of there is no cross air effect in with mesh condition. In another side without mesh condition have little amount of cross air effect which is known as cols inflow effect.

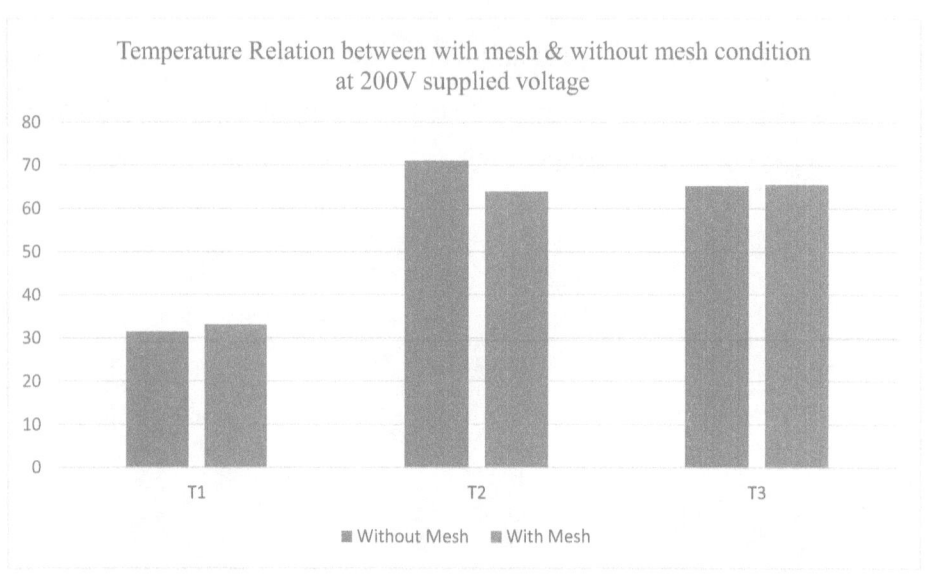

Figure 4.3: Temperature Relation between with mesh & without mesh condition at 200V supplied voltage

In figure 4.3,

For T1 temperature with mesh condition temperature is high cause is atmospheric temperature fall in without mesh condition.

For T2 temperature without mesh condition temperature is high cause of

For T3 temperature with mesh condition temperature is high cause of there is no cross air effect in with mesh condition. In another side without mesh condition have little amount of cross air effect which is known as cols inflow effect.

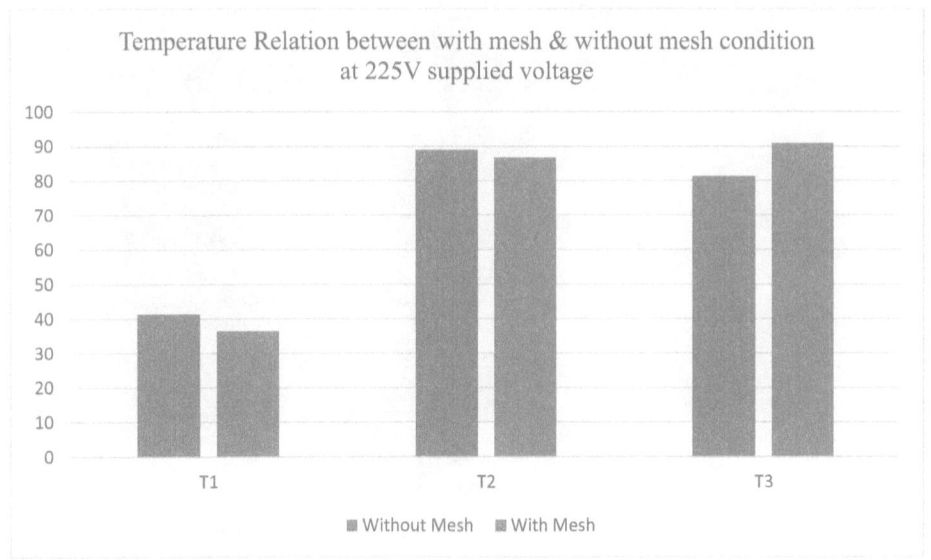

Figure 4.4: Temperature Relation between with mesh & without mesh condition at 225V supplied voltage

In figure 4.4,

For T1 temperature without mesh condition temperature is high cause is atmospheric temperature fall in with mesh condition.

For T2 temperature without mesh condition temperature is high cause of

For T3 temperature with mesh condition temperature is high cause of there is no cross air effect in with mesh condition. In another side without mesh condition have little amount of cross air effect which is known as cols inflow effect.

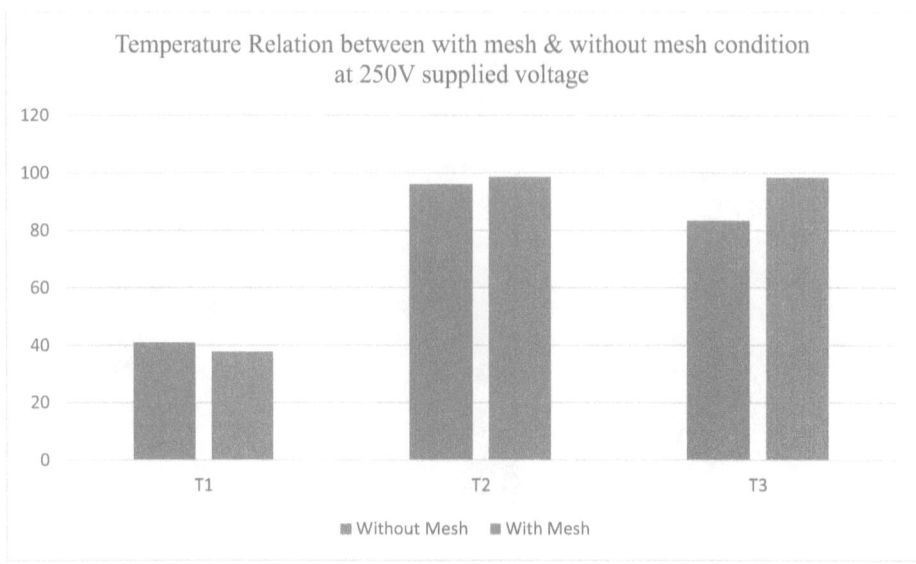

Figure 4.5: Temperature Relation between with mesh & without mesh condition at 250V supplied voltage

In figure 4.5,

For T1 temperature without mesh condition temperature is high cause is atmospheric temperature fall in with mesh condition.

For T2 temperature with mesh condition temperature is high cause of

For T3 temperature with mesh condition temperature is high cause of there is no cross air effect in with mesh condition. In another side without mesh condition have little amount of cross air effect which is known as cols inflow effect.

Heat Gain Ratio:

Table 4.12: Heat gain ratio

Voltage (V)	Density ($\Delta\rho$)	Diameter (d)	Area (A)	Velocity (v)	Mass (M)	Specific Heat Capacity (C_p)	Temperature Difference (ΔT)	Specific Heat (Q)	Power (P)	Heat Gain Ratio
			$\dfrac{\pi d^2}{4}$		($\Delta\rho Av$		T2-T1	$MC_p\Delta T$		$\dfrac{Q}{P}$
Volts	kg/m³	m	m²	m/s	Kg	kJ/kg.K	K	J/Kg	Watt	
Without Mesh										
150	0.079	0.3	7.07	1.004	0.561	1.0065	20.77	11.728	123.75	0.095
175	0.1	0.3	7.07	1.122	0.793	1.007	29.91	23.885	168.525	0.141
200	0.104	0.3	7.07	1.196	0.879	1.0077	39.61	35.085	220	0.16
225	0.126	0.3	7.07	1.411	1.257	1.0092	47.47	60.219	178.55	0.337
250	0.134	0.3	7.07	1.5	1.421	1.01	55.089	79.064	412.5	0.192
With Mesh										
150	0.070	0.3	7.07	1.17	0.579	1.006	18.34	10.683	123.75	0.086
175	0.102	0.3	7.07	1.31	0.945	1.0069	26.47	25.187	168.525	0.15
200	0.111	0.3	7.07	1.5	1.177	1.0072	30.88	36.607	220	0.167
225	0.171	0.3	7.07	1.22	1.475	1.009	50.3	74.86	178.55	0.420
250	0.185	0.3	7.07	1.4	1.831	1.01	60.86	112.549	412.5	0.273

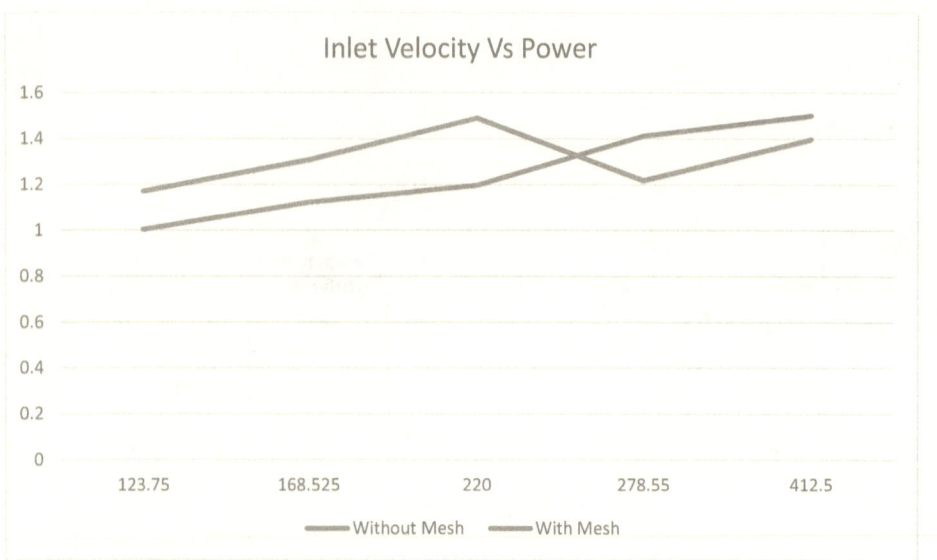

Figure 4.6: Relationship between inlet velocity & power

In figure 4.6, In without mesh condition we saw that when the power supply is increased then the air velocity which is flow in chimney is increased continuously.

In another side with mesh condition we saw that when the power supply is increased for a certain power air velocity is increased but at 220-278.55 W supplied power, the velocity is fallen down instantly. Then again when we supplied more than 278.55 W power, the inlet velocity of air is increased.

From figure 4.6, we able to saw that, in without mesh condition there is no effect of falling inlet velocity of air in chimney with increasing power. But in with mesh condition, we saw inlet air is increases then falling down & again it increases with power.

4.2 Discussion

For produce electricity without doing any damage of environment, solar chimney power plant is a good model for the world. Electricity is the basis of all the development of the world. But the amount of pollution we did with ourselves and earth is phenomenal. If we build this type of power plant for every city then it would be very effective and good for people and environment. Those area which has a good amount of sunlight effect (such as Asia, Africa & Australia) is suitable for this type of power plant. Total worldwide gross production of electricity in 2016 was 25,082TWh. Sources of electricity were coal and peat 38.3%, natural gas 23.1%, hydroelectric 16.6%, nuclear power 10.4%, oil 3.7%, solar/wind/geothermal/tidal/other 5.6%, biomass and waste 2.3%. If we increase the rate of electricity production using solar chimney power plant up to 20%, that would the greatest achievement for new generation and as well as the planet.

The advantages are given below:

a) Solar chimney power plants are particularly suitable for generation electricity in desert and sun rich wasteland.
b) It provides electricity 24 hours a day from solar energy alone.
c) No fuel is needed.
d) It is particularly reliable and a little trouble prone compared with other power plants.
e) The materials concrete, glass and steel necessary for the building of solar chimney power plants are everywhere in sufficient quantities.
f) No ecological harm & no consumption of resources.

The limitations are given below:

a) The cost of generation electricity from a solar chimney power plant is 5 times more from a gas turbine.
b) The structure itself is massive. So, large area is needed.

The applications are given below:

We can implement the project in the following sectors:

a) In power Plant
b) In Ventilation
c) In Industries as renewable energy source

CHAPTER 5

CONCLUSION & RECOMMENDATIONS

CHAPTER 5

CONCLUSION AND RECOMMANDATIONS

5.1 Conclusion

Solar chimney power stations could make important contributions to the energy supplies in Africa, Asia & Australia, because there is plenty of space & sunlight available. It is very important for the future because our resources are limited except our sun. Small solar fields can be integrated into fossil fuel power plants at relatively low costs.

With improvement of the cost-benefit ratio of STE, the solar share in hybrid solar/fossil power plants may increase to about 50 percent. Thermal energy storage will be able to further substitute for the need for a fossil back-up fuel. In the long run, baseload-operated solar thermal power plants without any fossil fuel addition are now technically proven.

In the Sunbelt of the world, solar thermal power is one of the candidates to provide a significant share of renewable clean energy needed in the future.

5.2 Recommendations

The project can be further improved by following ways-

- ▶ Can be use turbine at different place to get better efficiency.
- ▶ Design of chimney can be change for better efficiency.
- ▶ Gate can be installed at the bottom of chimney for turbine efficiency.
- ▶ More sharp data analysis can be done by installing mesh at different place.

REFERENCES

[1] P. J. Bansod1 et.al (2014) Solar Chimney Power Plant-A Review.

[2] Chi-Ming Chu,Md. Mizanur Rahman, Sivakumar Kumaresan et.al (2011) at Effect of cold inflow on chimney height of natural draft cooling towers.

[3] Xinping Zhou, Jiakuan Yang, Bo Xiao, Guoxiang Hou, Fang Xing et.al (2009) at Analysis of chimney height for solar chimney power plant.

[4] Hakim Semai, Amor Bouhdjar & Salah Larbi et.al (2016) at Canopy Slope Effect on the Performance of the Solar Chimney Power Plant.

[5] P. J. Bansod, S. B. Thakre, N. A Wankhade et.al (2014) at Solar Chimney Power Plant-A Review.

[6] Ajay Bejalwar, Pramod Belkhode et.al (2018) at Analysis of Experimental Setup of a small solar chimney power plant.

[7] Xinping Zhou, Fang Wanga, Reccab M. Ochieng et.al (2010) at A review of solar chimney power technology.

[8] Qingjun Liu, Fei Cao, Yanhua Liu, Tianyu Zhu and Deyou Liu et.al (2018) at Design and Simulation of a Solar Chimney PV/T Power Plant in Northwest China.

[9] Shinsuke Okada, Takanori Uchida, Takashi Karasudani, Takashi Karasudani et.al (2015) at Improvement in Solar Chimney Power Generation by Using a Diffuser Tower.

[10] AminMohamed El-Ghonemy et.al (2016) at Solar Chimney Power Plant with Collector.

[11] Amel Dhahri, Ahmed Omri et.al (2013) at A Review of solar Chimney Power Generation Technology.

[12] C. Chi-Ming Chu, R. Kwok-How Chu & M. M. Rahman et.al (2012) at Experimental study of cold inflow and its effect on draft of a chimney.

[13] Omer Khalil Ahmed, Abdullah Sabah Hussein et.al (2018) at Case study of new design of solar chimney.

[14] W. M.A Elmagid, I. Keppler et.al (2017) at Axial flow turbine for solar chimney.

www.ingramcontent.com/pod-product-compliance
Lightning Source LLC
Chambersburg PA
CBHW030533220526
45463CB00007B/2811